ASSUMPTIONS OF COSMOLOGY

ASSUMPTIONS OF COSMOLOGY

Data-Based Alternatives To Dark Energy And Dark Matter

WAYNE TAYLOR

LitPrime
"Your story is our priority"

LitPrime Solutions
21250 Hawthorne Blvd
Suite 500, Torrance, CA 90503
www.litprime.com
Phone: 1-800-981-9893

Published by LitPrime Solutions   03/28/2022

ISBN: 978-1-955944-70-0(sc)
ISBN: 978-1-955944-71-7(e)

Library of Congress Control Number: 2022904437

Contents

Foreword

Science is the best method we have for determining the truth. The method of science is to form hypotheses and assumptions, and then test them, Cosmology is the study of things that cannot be tested. Cosmology is not science. Cosmology consists entirely of untestable hypotheses and assumptions.

Inflation, Dark Energy, and Dark Matter started out as wildly imaginative hypotheses. These hypotheses have been sensationalized by the media, for so long, that they have gained general acceptance. None of the rational competing hypotheses have been popularized.

It is commonly believed that Cosmologists, being so intelligent, have eliminated all rational hypotheses, leaving the imaginative ones as the only possible explanations. It is assumed that there are no straight forward ways to replace these wildly imaginative hypotheses. This is saying that Cosmologists have proven a negative. It is impossible to prove any negative statement; such as, no alternatives exist. All of these wildly imaginative hypotheses have been brought together in the "Lambda CDM Standard Model Of The Early Universe" Based only on unsubstantiated guesswork, it is widely believed that 95 % of the Universe is composed of mysterious unknown matter and energy.

This paper offers simple alternatives, eliminating the need for the wildly imaginative hypotheses of Inflation, Dark Energy, and Dark Matter.

Hypotheses With Unanswered Questions

When hypotheses develop unanswered questions, cosmologists tend to come up with complicated alternatives. Imaginative hypotheses can turn out to be useful in unexpected ways. Even if the hypothesis proves to be untenable, the concepts and mathematics may be applicable in a future hypothesis.

Lorentz hypothesized that time and space may vary in a way that would make the ether undetectable. Even though physicists no longer believed that the ether was necessary for the propagation of electromagnetic waves, Einstein was able to use the concept and mathematics, from this discarded hypothesis, in his Special Theory of Relativity. Both hypothetical assumptions and empirical testing are necessary for the continued advancement of science.

Cosmologists are among the most intelligent people in the world. Their logic and reasoning are flawless. However, logic and reasoning are always based on assumptions. Often times, assumptions are taken for granted. It is important to give assumptions as much attention as we give to logic and reasoning. When logic defying assumptions are made, popular hypotheses can lead to logic defying conclusions.

The "standard model of the early Universe" is only a hypothesis based upon the following unproven assumptions.

> 1   The Singularity created all of the matter and energy in the Universe.

2   We can have no idea what the Universe is like outside of the Visible Universe, VU. Therefore, the VU is the entire Universe.

3   The Singularity created time and space.

4   The CMB is a picture of the Big Bang

The following assumptions have recently been added.

5   Dark Energy exists

6   Dark Matter exists.

7   Inflation existed

The name has also been changed to "The Lambda CDM Standard Model Of The Early Universe". Using "Standard Model" as part of the name is an attempt to add legitimacy to this wildly imaginative hypothesis.

When assumptions are changed, the logic and reasoning which follow will also be changed and the Universe will become more understandable in terms of rational results.

Occam's Razor

A heuristic is a non-rigorous rule that tends to lead to correct solutions of problems. Occam's razor is a heuristic that states "the simplest solution that answers all of the questions, has the highest probability of being correct". A good heuristic for solving the problems of cosmology is; "Stick to known facts as much as possible, and use Occam's razor on any and all assumptions".

The Visible Universe, VU, is located at a random point within the entire Universe. Objects outside of the VU are receding away so fast that their light will never reach us.

A Simple Model Of The Universe

By careful scrutiny of assumptions, we can create a model of the Universe where; there is no Dark Energy, there is no Dark Matter, and there is no Inflation.

Popular cosmological hypotheses claim that we can know nothing of the Universe outside of our VU, Visible Universe, since we are unable to observe it. Since we can know nothing, then universe may be infinite. If the universe is infinite, then anything can be possible. Wildly imaginative hypotheses flourish such as the multiverse, the bubble Universes and, the extra dimensional Universes. We are drowning in a flood of hypotheses of what the Universe might be like outside of the VU.

With the following assumptions, we can logically demonstrate that the Universe is a much simpler place.

1   The Universe was created by some type of Big Bang, or violent event, not necessarily a Singularity.

2   The Universe is an expanded version of the Visible Universe, VU. Remembering Occam's razor, this is the simplest assumption we can make about the Universe.

3   The VU is located at some random point within the expanding cloud of the Universe.

4   Light from beyond the VU can never reach us, see the section "CMB Not From Big Bang". The CMB is a picture of the outer edge of the VU. It is not a picture of the Big Bang, or the beginning of the Universe.

The Expanding Universe

Before the launch of the Hubble Space Telescope, HST, the Universe was commonly believed to be infinite. Observations from the HST revealed that we could see nothing beyond a cosmic horizon of about 14 billion light years in all directions. This observable sphere was termed the Visible Universe, or the VU.

Many theories have assumed that the VU is the entire Universe. This mistaken assumption has led to many wrong conclusions. The edge of the VU is the Cosmic Horizon.

We can extrapolate, in accordance with the Hubble constant that, the Universe beyond the cosmic horizon is receding away from us faster than the speed of light. See the appendix. Many scientists recoil from this idea, hiding behind Einstein's theory of relativity. However, the theory of relativity requires that two objects must share a common reference frame to satisfy the restriction that nothing can travel faster than the speed of light. Objects outside of the VU cannot share a common reference frame with our Solar system, which is at the center of our VU. Interestingly, objects just inside the edge of our VU will be able to observe objects just outside of the edge of our VU expanding away at a speed predicted by the Hubble constant, based upon the distance between the objects.

The Dark Energy Alternative

In 1998 the Supernova Cosmology Project and the High Z Supernova Search both released the results of their measurements of the luminosity of type 1A supernovae. The key finding was that the supernovae were fainter than expected. They cautiously suggested the hypothesis of an accelerating rate of expansion of the Universe and thus, Dark Energy might be true, resulting in a Nobel Prize in 2011. *[Adam Riess et al. (Supernova search Team) 1998). "Observational evidence from supernovae for an accelerating universe and a cosmological constant"]*. Dark Energy has become an accepted assumption among scientists, even though the assumption of Dark Energy is based upon the cautiously worded suggestion of the supernovae studies.

Measuring the luminosity of the supernovae is an independent method of confirming the validity of the redshift measurements. Redshift measurements have always been confirmed by luminosity measurements for distances less than about a billion light years. The 1998 luminosity measurements were the first attempts to confirm redshifts of objects at distances greater than a billion light years.

Since the supernovae were fainter than expected, the luminosity measurements did not agree with the redshift measurements. Being fainter than expected, means the supernovae were further away than expected. This was interpreted to mean the supernovae were traveling faster than the redshift predicted. Therefore, they were traveling away from us, in the past, at a rate faster than predicted by the Hubble Constant, H_0, which is currently estimated to be 67.80 (Km/s)/Mpc.

This has been interpreted to mean that the rate of expansion of the Universe **is** accelerating. The correct interpretation should be that the rate of expansion is decelerating. They had to assume that there was no dust obscuring the luminosity of any of the supernovae. They also wrongly interpreted a faster expansion in the past as meaning the rate of expansion is accelerating, rather than decelerating.

The Nobel Prize Is not given for theories or hypotheses: it is only given for discoveries, experiments, or observations. The scientists, who won the Nobel Prize, for precise observations of the luminosity of distant supernovae. Their Nobel Prize had nothing to do with their hypothesis that the expansion rate of the Universe is accelerating. They would just as easily have won the Nobel Prize if they had concluded that the expansion rate was decelerating.

No Limit To The Mass Of Universe

The "standard model of the early universe' hypothesizes that the original baryon to photon ratio sets a limit to the amount of matter in the universe. The estimated mass of light nuclei created by the Big Bang is claimed to not add up to enough matter. They estimate a current baryon to photon ratio, based on the abundance of deuterium and of photons coming from the CMB.

It is claimed that the number of photons coming from the CMB is unchanged from the time of the Big Bang. It is a common misconception that we can now observe the universe all of the way back to the time of recombination, shortly after the Big Bang. We can really only observe the Visible Universe, VU, out to the cosmic horizon. The current CMB cannot tell us what the photon count was at the time of recombination. So, we do not know what the baryon to photon ratio originally was, and we definitely cannot set a limit to the mass of the universe.

The CMB power spectrum graph is hypothesized to be another way to determine the original baryon to photon ratio. WMAP and BICEP experiments have observed variations of temperature in the CMB. These temperature ranges can be used to plot a graph called the CMB power spectrum. The temperature peaks of this graph are supposed to represent the baryon to photon ratio at the time of recombination.

The CMB is a map of the relative density of photons of the most distant sources of light near the cosmic horizon. This is not light from the beginning of the Universe. Rather, it is light form the edge of the VU. It is a picture of the relative density of galaxies near the cosmic horizon of the VU at a time about 14 billion years in the past. The CMB power spectrum graph may possibly tell us what the baryon to photon ratio was 14 billion years ago. It cannot tell us what the ratio was at the time of recombination. Therefore, we do not know what the mass of the Universe should be.

The Dark Matter hypothesis Alternative

One reason for a hypothesis of Dark Matter is that the outer portions of galaxies and clusters should be rotating slower than the inner portions, based upon the mass indicated by light imaging. We informally use light imaging to represent all forms of electromagnetic imaging. Another reason for a hypothesis of Dark Matter is that light imaging indicates less mass in galaxy clusters and super clusters than does gravitational lensing. [Sean Carroll, Ph.D., Cal Tech, 2007, The Teaching Company, *Dark Matter, Dark Energy: The Dark Side of the Universe*, Guidebook Part 2 page 46] The missing mass could readily be explained by an equivalent mass of Black Holes, faint stars, rogue planets, asteroids, comets, gas, dust, and other forms of matter that would not be telescopically visible

There are two ways to account for enough mass in the Universe without resorting to a hypothesis of Dark Matter. The first way is to realize that we don't have a limit to the mass of the Universe because we don't know the original ratio of baryons to photons. The second way to account for enough mass in the Universe is to note that the standard model of the early Universe is based on the assumption that the entire Universe was created from a Singularity, which would have had no primordial (preexisting) Black Holes. If we assume that the early Universe started from an exploding Big Crunch. There could well have been primordial Black Holes, thus, allowing for the observed mass of the Universe.

Dark Matter is no longer a necessary assumption, since we

have a more viable alternative to the Singularity assumption. There could easily have been enough matter from exploding Black Holes to account for the mass of the universe.

Before the hypothesis of the "mysterious" Dark Matter, the dark matter of galaxies was thought to consist of black holes and burned-out stars in the galactic halos.

(Genesis: the origins of man and the Universe By John Gribbin)

Inflation Hypothesis Alternative

Without the Horizon problem, one of the reasons for a hypothesis of Inflation has vanished. The inflationary hypothesis was developed in the 1980s by physicists Alan Guth and Andrei Linde. *[Chapter 17 of Peebles (1993)]* The other reason for Inflation, the homogeneity and isometry of the Universe, could be removed by a momentary deactivation of the Higgs boson during the Big Bang. A momentary lack of mass, and thus, lack of gravity following the Big Bang is more feasible than faster than light inflation. Perhaps a theory of slower than light, mass free, inflation could be developed based upon the Higgs boson.

Another alternative to Inflation could be the few hundred thousand years, dominated by the Compton Effect, before the epoch of recombination, or first light. The high density of photons (gamma rays), interacting with electrons, inhibited the formation of atoms. Without the electron bonding of atoms, there was no clumping of matter during this period. At the atomic level gravity is insignificant compared to the electromagnetic force. The positive charge of protons would have a much greater repulsive force than the meager attraction of gravity. Random groupings of baryons (protons and neutrons) could form temporary gravitational clusters of matter. This could well have generated the baryon acoustic oscillations. Without electron bonding these groupings would be randomly colliding and rebounding, Temporary clusters of matter would be forming and dissolving. This could be considered as a non-clumping inflationary period. Therefore, there is no need for a hypothesis of faster-than-light inflation.

Large Scale Structure Of The Universe

Dark Matter is hypothesized to be necessary for the formation of the galaxies. It is thought that gas clouds large enough to form galaxies would not have had time to accumulate, given the rate of expansion of the Hubble constant. This is a reasonable assumption. However, there would have been time for many individual stars to form. Given the high density of the early Universe, many of these stars would have been massive and short lived. The supernovae of these earliest massive stars would expel great amounts of matter. The result of this expulsion could will have created the great voids, and compressed the surrounding matter into what we call the large scale structure of the Universe. More stars could then form and go nova within the compressed matter. This would explain the large voids and the bubbles and foam structure.

The Singularity Hypothesis

A PBS TV show, about the Big Bang stated that over half of the scientists, at the Perimeter Institute, no longer believed that a Singularity caused the Big Bang. They believed that the Universe was created by some type of violent event, not necessarily by a Singularity.

Like most people of his time Lemaitre was fascinated by the discovery of radioactivity. He hypothesized that the creation of the Universe might possibly be analogous to the radioactive decay of some very heavy element. In 1931 Lemaitre suggested that the evident expansion of the Universe, if projected back in time, meant that the further in the past the smaller the Universe was; until at some finite time in the past all the mass of the Universe was concentrated into a single point, a "primeval atom" where and when the fabric of time and space came into existence. This has slowly changed over the years from being thought of as a wildly imaginative hypothesis into being widely accepted, even though; we have no conceivable evidence that such a thing is even possible. Furthermore, there is no logical reason why a rewind of the Universe would have to continue all the way back to the size of an infinitesimal point. While we no longer believe in a "Primordia Atom": we have hypothesized the ideas of a Singularity, or possibly large vacuum fluctuations. We do not have any evidence or observations singularities or large scale vacuum fluctuations. The assumption of a Big Bang from an exploding Big Crunch is more consistent with experimentally verifiable data than the assumption of a Big Bang from a Singularity or massive vacuum fluctuations.

The Big Crunch

Cosmologists agree that the Universe was created by some type of violent event, but most of them no longer believe that the entire Universe was created from a Singularity.

A viable alternative to the current Big Bang from a Singularity hypothesis would be a Big Bang resulting from an explosion of multiple Black Holes in the late stages of a Big Crunch. In a Big Crunch most of the stars will have burned out and consolidated with other stars into Black Holes, and the Black Holes will slowly gravitate towards each other. A large enough cluster of Black Holes could contain the mass of our Universe. An explosion of such a cluster could account for the creation of the matter and energy in the Universe equally as well as the traditional Big Bang hypothesis.

The calculations for an exploding Big Crunch would be different from the calculations of the early universe created from an idealized Singularity. A chain reaction of exploding Black Holes would take place over an extended period of time and, over extended distances in space. Black Holes distant from the central cluster could have survived as primordial Black Holes,

We do not know why a Big Crunch would explode, any more than we know why a Singularity would explode. However, the assumption of an exploding Big Crunch relies more on experimentally proven facts and has fewer unanswered questions then the assumption of an exploding Singularity.

Exploding Big Crunch Hypotheses

An older Big Bang hypothesis states that a critical mass is reached when the Universe finally contracts into a single Black Hole, and a Big Bang is triggered. This older hypothesis assumes the critical mass for a Big Bang is the entire mass of the Universe. What if the critical mass is smaller that this? Then whenever a Black Hole reaches the critical mass, it would cause a Big Bang with many distant unexploded primordial Black Holes.

A second alternative would be finding plausible circumstances to make Black Holes explode. For example, a very large supernova exploding close to a cluster of Black Holes ranging in size from minimal mass to super giants. We could throw in some neutron stars and give everything a rapid spinning motion for good measure. A minimal mass Black Hole could possibly be affected by the nearby explosion of a gigantic supernova. Here we can assume that a Black Hole has shrunken to its minimal mass by Hawking radiation, and is barely able to hold itself together.

Another plausible circumstance to make Black Holes explode would be a grazing collision of rapidly spinning Black Holes. This would be an example of an immovable object being hit by an unstoppable force.

Age Of The Universe

Objects farther than 14.42 billion light years from us would be receding away from us faster than the speed of light, as explained in the earlier section "The Expanding Universe". This distance is calculated in the appendix.

We can assume our sphere of the VU is at a random point within the expanding cloud of the Universe. As a type of mathematical thought experiment, we can imagine rewinding backwards to the beginning of the Universe. Every point, including the center of VU, would eventually coincide with the center of the Universe, theoretically.

Since both VU and the Universe will rewind to the same center point in the same time interval, this thought experiment reveals that VU and the Universe also expanded to their respective sizes in the same time interval.

If we think about it, this is the age of VU at a time 14.42 billion years in the past; since, it would take that long for the most distant light to reach us. The VU and the Universe would have been about 14.42 billion years old at a time 14.42 billion years in the past. This would place the age of the Universe at about 28 billion years.

NOTE this estimate is based on the current Hubble constant, H_o. If the rate of expansion is decelerating then the age of the universe could be less than 28 billion years.

The age of the universe is also estimated by the age of the

oldest stars. Another method of estimating the age of the universe is by measuring the ratio of radioactive isotopes in the oldest stars. The isotope ratios can be measured for thorium and uranium/ also for rhenium and osmium. We can only observe the oldest stars in the VU. We cannot observe the oldest stars in the Universe. Likewise, we can only measure the ratio of radioactive isotopes in the oldest stars in the VU. Since our observations are limited to the VU, we really cannot estimate the age of the Universe.

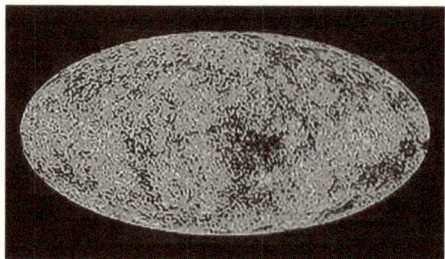

The Cosmic Microwave Background, CMB, is a picture of the light coming from the edge of the Visible Universe. The light from the rest of the Universe will never reach us. Thus, the CMB is not a picture of the Big Bang. It is a picture of the light emitting bodies at the edge of the VU.

CMB Not From Big Bang

Simply put, the CMB, Cosmic Microwave Background, is a picture of photons at a certain frequency. Photons are the carrier particles for all frequencies of the electromagnetic spectrum, from Gamma rays down through heat radiation. Talking about the temperature, or heat radiation, is just another way of talking about the frequency of photons. The temperature of the CMB is 2.725 degrees K. This is also said to be the temperature of empty space, since the CMB goes everywhere in space.

The CMB is said to be the oldest light in the Universe, coming from the epoch of recombination, just 380,000 years after the Big Bang. This interpretation is incorrect. The CMB is only the oldest light in the Visible Universe. The cosmic horizon is the outer edge of our VU. The CMB consists of light coming from near the cosmic horizon, which is nowhere near the big bang or the epoch of recombination.

The cosmic horizon is the distance at which the Universe is expanding away from us at the speed of light. At distances farther away from us than the cosmic horizon, the Universe is expanding away from us faster than the speed of light. So, the light coming from beyond the cosmic horizon will never reach us.

The correct interpretation is that the CMB is a picture of the cosmic horizon, the oldest part of our VU. This limits our observations to the VU, which is as it should be

Time And Space Were Not Created

There is no logical reason why time and space need to have been created. Lematre was the first to propose that time and space were created, and this hypothesis has been adopted in both the hypothesis of inflation and the hypothesis of the standard model of the early universe. Lematre proposed that time and space were created by the "Primevil Atom". The hypothesis of Inflation, by Alan Guth, predicts that time and space were created by a singularity. The hypothesis of Inflation and the hypothesis off "the standard model of the early universe: both assume that only space were expanding faster than light. Therefore. No objects would need to have traveled faster than light. These are highly complex assumptions. We can eliminate the complexity by assuming that a finite Universe is expanding within an infinite space. The dots on a balloon analogy explains the faster than light expansion, outside of the VU.

Einstein's theory of General Relativity predicts that time and space will shrink down to nothing at the center of a Black Hole, or a Singularity. Black Holes are said to be closed off from time and space, yet time and space still exist all around them. If a Black hole were somehow made to explode, it would not be creating time and space

Horizon Problem Explained

The Horizon Problem states that the Universe has not existed long enough for information (matter, energy) to travel through its vast distances. *[First identified in the late 1960s, primarily by Charles Meisner]* We are at the center of our VU with a radius of approximately 14.42 billion light years (this number is estimated in the appendix at the end of the paper). This would give us a spherical diameter of 28.8 billion light years. This seems to be an impossible distance to travel in the time since the Big Bang, assuming that nothing can travel faster than light.

If we picture the very early Universe as an expanding cloud of matter and energy, we could imagine placing an observer at the center of the Big Bang. All matter and energy would be expanding outward at or below the speed of light. Using the dots on a balloon analogy, the dots furthest away from the observer would be moving away faster than the closer dots. After a certain point, the farthest dots from the observer would be receding at a rate faster than the speed of light. Keeping in mind that the Universe is much larger than the VU, the great distances and speeds noted in the Horizon problem are easily explained by the dots on a balloon analogy. Therefore, there is no Horizon problem

Summary

A good heuristic for solving the problems of cosmology is; "Stick to experiments and observations as much as possible, and use Occam's razor on any and all assumptions". This paper makes the following assumptions, allowing the Universe to become a much more reasonable place.

1   The Universe was created by some type of Big Bang, or violent event, not necessarily a Singularity.

2   The Universe is an expanded version of the VU. Remembering Occam's razor, this is the simplest assumption we can make about the Universe.

3   The VU is located at some random point within the expanding cloud of the Universe.

4   The CMB is not a picture of the early Universe, since we cannot observe anything outside of the VU.

Using these assumptions, we can come to the following conclusions.

The CMB is a picture of the density of the galaxies near the edge of the VU.

The Singularity hypotheses assumes that we know the original baryon to photon ratio, and thus, we have a limit to the mass of the universe. We have shown that this ratio cannot be determined from inside the VU. Taking a pragmatic approach, we can say that the Singularity hypothesis has been contradicted. Since this hypothesis has predicted less mass than has been observed in the Universe.

The existence of dark matter is shown to be unnecessary since the mass limiting baryon-photon ratio is unknown at the time of recombination, shortly after the Big Bang.

The fainter than expected supernovae has been interpreted as suggesting an accelerating rate of expansion of the Universe. This, the possibility of dark energy is hypothesized. The correct interpretation implies a decelerating rate of expansion of the Universe. Therefore, Dark Energy does not exist. Since different supernovae have variations in their predicted luminosity, it is probable that dust clouds are obscuring our vies of the distant supernovae. Then, there would be no change in the rate of expansion.

By using the dots on a balloon analogy, we can show that some parts of the Universe will be traveling away from each other faster than the speed of light. The Universe has had a constant, or possibly decreasing, rate of expansion since its creation. Thus, there is no Horizon problem.

Without the Horizon problem and with the possibility that the mass-providing Higgs Boson might have been temporarily deactivated following the Big Bang; and also considering the non-clumping of matter for several hundred thousand years due to the Compton Effect previous to the epoch of recombination: we have two types of slower than light inflation. There is no need for a hypothesis of faster-than-light Inflation.

Expansion of the VU to a radius of 14.42 billion light years, at a point in time 14.42 billion years in the past: places the age of the Universe at 28.8 billion years.

NOTE this estimate is based on the constant rate of expansion of the Hubble constant, H_o. If the rate of expansion is decelerating then this number will be reduced in size.

Appendix

EXPANSION RATE CALCULATION

NOTE   These calculations assume that the Hubble constant has had the same value since the origin of the Universe. The fainter than expected supernovae indicate that the expansion rate of the Universe is decelerating. Therefore, the Hubble constant may become the Hubble variable. Beyond a billion light years, distances and speeds will be under estimated by the current Hubble constant.

Rate Of Expansion

With our newest telescopes, we see approximately 14 billion light years in all of the directions of a sphere with ourselves at the center. Other names for the Visible Universe, VU, are the Observable Universe and the Hubble Sphere. We can detect nothing beyond this distance. Names for the surface of the sphere expanding away from us at the speed of light are, the Surface of the Hubble Sphere, the Cosmic Horizon, or just the Horizon.

We will look at a conversion formula yielding answers in billions of light years. When looking at distant objects, this is a more convenient unit of measure than (Km/s)/Mpc (kilometers per second per megaparsec).

We can confirm the assumption that we are in a sphere of the VU, with calculations using H_o, the Hubble Constant. The formula we use will show that the Universe is expanding faster than the speed of light at distances beyond 14.42 billion light years. We are unable to see objects which are receding away from us faster than the speed of light, because the light can never reach us.

The Hubble constant H_o, is approximately 67.80 + or − 0.77 kilometers per second for every mega parsec of distance. The margin of error is 0.77 / 67.8 = 1%. We can find R_H (the radius of VU) by calculating the distance at which objects would be traveling away from us at the speed of light; using the following conversion formula to calculate the radius, R_H, to the Cosmic Horizon

$R_H = c / (H_o*Ly)$ or
$R_H = c * Mpc / (67.8*Ly)$ or
$R_H = 14,422,034,697.57$ or
$R_H \sim 14.42$ billion light years, where
NOTE The Hubble constant is a fraction (Km/s)/Mpc. When dividing by a fraction, invert and multiply.

	Hubble constant	
H_o	~ 67.80	(Km/s)/Mpc
	Speed of light	
c	= 299792.458	Km/s
	Light year	
Ly	= 9.4605284×10^{12} Km	
	Mega parsec	
Mpc	= $3.08568025 \times 10^{19}$ Km	

Allowing for a 1% margin of error, R_H = 14.42 + or - 0.14 billion light years. Since H_o is a constant in a proportional relationship, we can derive the following formula from the equivalent ratios.

(1) $d / R_H \sim v/c$
In other words:
distance / radius of VU $\sim$ velocity / c

v/c is the apparent velocity of the object, as a fraction of the speed of light. D is the distance of the object in billions of light years. R_H (~ 14.42) is the radius of VU in billions of light years, which can be recalculated for improved values of H_o using the previous conversion formula.

Just as the Hubble constant, H_o, says that the Universe is expanding at a rate of 67.8 kilometers per second for every mega parsec of distance, the formula $v/c \sim d/14.42$ says that the Universe is expanding at a rate of c for every 14.42 billion light years of distance. It's just a scaled up fraction equal to the Hubble constant.

67.8 (Km/s)/Mpc = c / 14.42 billion light years.

Using formula (1), an object at a distance of 8 billion light years should be expanding away from us with a velocity of

$v/c \sim 8 / 14.42 \sim .55c$.

The galaxy at a distance of 13 billion light years should be expanding away from us with a velocity of

$v/c \sim 13 / 14.42 \sim .9c$.

An object on the outer surface of VU should be expanding away from us with a velocity of

$v/c \sim 14.42 / 14.42 \sim c$.

An object at a distance of 20 billion light years should be

expanding away from us faster than the speed of light with a velocity of

$$v/c \sim 20 / 14.42 \sim 1.39c$$

The most distant object we can view is a galaxy 13 billion light years away, traveling at very near the speed of light. An object traveling away from us at the speed of light would be just a little farther, at a distance of 14.42 billion light years. These are estimates using the above formulas.

Groups, clusters, and clusters of clusters (also called super clusters) of galaxies are each bound together by gravity. The Galaxies within a group, or cluster, or super cluster will be held together by gravity as the rest of the Universe is expanding. Centaurus, the nearest super cluster is at a distance of roughly 140 million light years. This would make its rate of expansion

$$v/c \sim .14 / 14.42 \sim .01c.$$

We can expect a slow rate of expansion, between neighboring super clusters, anywhere in the Universe, due to the dots on a balloon analogy.

Easy Distance Approximations From Z

For objects outside of our own Virgo super cluster we can use the following two formulas to approximate distance from the redshift.

The velocity of distant objects can be calculated from the redshift using the following velocity formula, where z represents the redshift.

(2) $v/c = ((z+1)^2 - 1) / ((z+1)^2 + 1)$

Solving the distance formula (1) from the previous section for the variable d yields the following distance formula.

(1) $d / R_H \sim v/c$
$d \sim R_H(v/c)$
By substituting 14.42 for R_H

(3) $d \sim 14.42(v/c)$

A spreadsheet can be easily set up using formula (2) to calculate v/c given z. This answer can then be multiplied by 14.42 to give the distance d, using formula (3). These two calculations can then be copied down two columns of a spreadsheet to give the distances for many different values of z, or redshifts.

The unit of distance is in billions of light years. There is usually a margin of error assigned to each redshift, and the distance formula (3) has the same margin of error as the Hubble Constant, Ho, which is 1%.

www.ingramcontent.com/pod-product-compliance
Lightning Source LLC
Chambersburg PA
CBHW031238120626
46545CB00003B/1185